WEAPONS

Peggy J. Parks

BLACKBIRCH PRESS
An imprint of Thomson Gale, a part of The Thomson Corporation

Detroit • New York • San Francisco • San Diego • New Haven, Conn. • Waterville, Maine • London • Munich

© 2005 Thomson Gale, a part of The Thomson Corporation.

Thomson and Star Logo are trademarks and Gale and Blackbirch Press are registered trademarks used herein under license.

For more information, contact
The Gale Group, Inc.
27500 Drake Rd.
Farmington Hills, MI 48331-3535
Or you can visit our Internet site at http://www.gale.com

ALL RIGHTS RESERVED
No part of this work covered by the copyright hereon may be reproduced or used in any form or by any means—graphic, electronic, or mechanical, including photocopying, recording, taping, Web distribution or information storage retrieval systems—without the written permission of the publisher.

Every effort has been made to trace the owners of copyrighted material.

Picture Credits:
Cover, © Bettmann/CORBIS (top); © Photri/SYGMA/CORBIS, (bottom)
© The Art Archive, 4(top),7(bottom left),7-8(arrowheads),10, 11(top),14
© Bridgeman Art Library, 19(bottom)
© Archivo Iconografico,S.A./CORBIS, 9
© Anthony Bannister,Gallo Images/CORBIS, 6(top)
© Bettmann/CORBIS, 9, 12, 15(bottom),17(top), 18(bottom), 22, 23, 24, 27(top)
© CDC/PHIL/CORBIS, 28
© Hulton-Deutsch Collection/CORBIS, 20(top), 21(bottom)
© The Mariner's Museum/CORBIS, 19(top)
© The Philadelphia Museum of Art/CORBIS, 15(top)
© Schenectady Museum,Hall of Electrical History/CORBIS, 20(bottom)
© Stapleton Collection/CORBIS, 21(top)
© CORBIS, 16, 25
© Hulton Archive by Getty Images, 8, 13, 18(top)
© Bloomberg News/Landov, 31(top)
© UPI/Landov, 31(bottom)
© Landov, 29(both), 30
National Archives, 26
Smithsonian Institution, 27(bottom)

LIBRARY OF CONGRESS CATALOGING-IN-PUBLICATION DATA

Parks, Peggy J., 1951–
 Weapons / Peggy J. Parks
  p. cm. — (Yesterday and today)
 Includes bibliographical references and index.
 ISBN 1-56711-836-4 (hardcover : alk. paper)
 1. Weapons. I. Johnson, Bob G. II. Title. III. Series: Yesterday and today.
 U800.P37 2005
 623.4'09—dc22

Printed in the United States of America
10 9 8 7 6 5 4 3 2 1

Table of Contents

The First Weapons	4
The Bow and Arrow	6
Weapons from Metal	8
Siege Engines	10
The Power of Gunpowder	12
Handguns and Rifles	14
Powerful Explosives	16
Weapons at Sea	18
Armored Vehicles	20
War Planes	22
Chemical Weapons	24
Nuclear Weapons	26
Biological Weapons	28
Weapons in a High-Tech World	30
Glossary	32
For More Information	32
Index	32
About the Authors	32

The First Weapons

Prehistoric hunters spear a mammoth. The spear was one of the first man-made weapons.

Humans who lived during prehistoric times did not have weapons like those of today. When they hunted game or fought against their enemies, they used whatever they could find. They likely grabbed rocks and threw them or used tree branches as clubs. They had one goal: survival.

Eventually humans began to make weapons. One of the earliest weapons was the spear. They would break a strong branch off a tree and scrape one end against a boulder until it was sharp. Then they would harden the sharpened end by burning it. Later, people made spearheads from pointed stones or animal bone and tied them to the end of a spear.

This made the weapons even more deadly. Spears could be thrown through the air or used in hand-to-hand combat.

Another ancient weapon was the sling. This weapon allowed stones to be hurled faster and farther than they could be thrown. Slings were made of long strips of animal hide, vine, or twisted human hair. They had a pouch in the center to hold a stone and a finger loop at one end. The slinger pushed a finger through the loop, gripped the loose end in the palm of the hand, and loaded a stone in the pouch with the other hand. Then the slinger raised the sling above the head and whirled it around and around. When ready to shoot, the slinger let go of the loose end and sent the stone flying through the air toward the target.

Ancient humans used weapons that were simple by today's standards. For the people who relied on those weapons for survival, however, they could mean the difference between life and death.

> ## Ancient Throwing Sticks
>
> A hunting weapon known as the kylie was used by aboriginal peoples in ancient Australia. Kylies were heavy, curved sticks that were thrown to stun or kill prey. Over time the sticks were made lighter and more curved, so when thrown they would return to the thrower. They eventually came to be known as boomerangs. Most historians believe boomerangs were likely used for sport rather than hunting.

Later hunters added a stone or bone spearhead to make the spear a deadlier weapon.

As the first long-range weapon, the bow and arrow had enough force to kill large animals.

The Bow and Arrow

With slings, ancient humans learned that weapons were more powerful when they were shot through the air. This led to the development of the bow and arrow, the first long-range weapon. Historians believe the bow and arrow originated around 25,000 B.C., or possibly earlier. By pulling back on a bowstring and letting it go, a shooter could propel arrows great distances. Also, arrows shot from a bow had great penetrating force—enough to kill even enormous animals such as buffalo.

Bows were made of whatever strong, flexible material was available. Some were made of hickory or other types of wood, while others were made of bamboo or animal horns. Bowstrings, which attached to both ends of a bow, were usually made of rawhide or sinew. Some were made of animal hair or plant fibers that were twisted for added strength.

Arrows were made of wood or bamboo that was carved to be as straight as possible. At first, arrow tips were sharpened and then hardened over fire. Eventually, people attached separate arrowheads, which varied based on the individual culture or tribe. Some were carved from animal bone, but most were made from a stone known as flint. The flint was chipped to make a razor-sharp point, and

The Deadliest Bow and Arrow

During the Middle Ages, longbows were commonly used by the British army. They were made of wood and measured between 5 and 6 feet (1.5 and 1.8m) long. Longbows were powerful weapons, capable of hitting targets hundreds of yards away. They could even penetrate solid steel armor. Because they could shoot ten to twelve arrows a minute, they are sometimes called medieval machine guns.

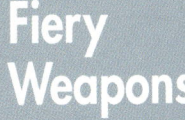

the arrowheads were tied to the end of the arrows. A small fin of bird feathers was added to the other end to help guide the arrows in flight.

The bow and arrow was developed thousands of years ago. Whether it was used to shoot game or to ward off dangerous enemies, people considered it a necessary and valuable weapon.

Fiery Weapons

Some ancient armies used flaming arrows. Just before letting go of a bowstring, soldiers lit the arrow tips on fire. When the arrows struck the target, the flames quickly spread.

An arrow shot from a longbow could penetrate steel armor.

Prehistory

500 B.C.

100 B.C.

A.D. 100

200

500

1000

1200

1300

1400

1500

1600

1700

1800

1900

2000

2100

7

Weapons from Metal

Men make weapons during the Bronze Age. Melting tin with copper made bronze, a stronger metal.

The flint used to make arrowheads was also used for other weapons. For instance, daggers were often made of flint, as were axes. There was a disadvantage to using flint, however. It was a brittle stone that often broke apart on impact.

Around 5000 B.C., humans discovered copper. Copper and other metals are found inside rock (known as ore), so the discovery was likely accidental. People probably thought copper ore was just another type of rock. Once they learned that heating copper ore caused it to melt, they began using it to make weapons. They made molds out of stone or clay and poured the molten metal into them. This process is known as casting. Humans used this method to make copper axes, spearheads, arrowheads, and knives.

Copper is a soft metal, so weapons made with it were not very sturdy. People realized they could make it more durable by combining it with other metals. For instance, they mixed melted copper with small amounts of tin to

Metal Suits

In addition to its use in weapons, metal was also used to make protective clothing. During medieval times, knights wore a type of body armor called chain mail. These suits were made with thousands of iron rings tightly linked to form metal fabric. There was a flaw, however. Chain-mail suits had holes, so arrows could penetrate them. To better protect knights in battle, solid steel armor eventually replaced chain-mail suits.

A medieval knight wears a chain-mail suit made of thousands of linked iron rings.

create bronze. Beginning about 3000 B.C., bronze was used to make nearly all weapons, so the period came to be known as the Bronze Age. Castings grew more advanced, and simple bronze daggers became more fancy. People of wealth and high status carried daggers with handles decorated with gold or elegantly carved ivory. Bronze was also used to make swords that were 2 or 3 feet (61 or 91cm) long.

About the eleventh century B.C., people discovered how to use an even harder metal called iron. From that point on, both bronze and iron were used to make weapons.

Whether metal was used to make simple spearheads or elaborate swords, it changed weapons forever. By casting metal, humans could make weapons that were sharper, stronger, and sturdier than anything they had made before.

These Bronze Age swords found in Spain show the weapons' deadly points.

Prehistory

500 B.C.

100 B.C.

A.D. 100

200

500

1000

1200

1300

1400

1500

1600

1700

1800

1900

2000

2100

Siege Engines

As people grew more skilled at making weapons, they began to build war machines. These were called siege engines because they were designed to destroy cities or other areas that were under siege, or surrounded by soldiers.

The earliest known siege engine was the battering ram, which was developed about 900 B.C. Battering rams were used to break down the heavy doors and walls of castles or other structures. They were usually made of tree trunks and many had iron heads shaped like a ram on the battering end. Battering rams were either held by teams of soldiers or mounted on sturdy wooden frames on wheels. Others were hung from support frames with ropes so they could be repeatedly swung against a target.

There were also siege engines designed to hurl objects such as large stones, called projectiles, great distances through the air. One example was the catapult, which was invented about 400 B.C. A catapult was a wooden machine with a vertically swinging arm and tension created by ropes. A catapult's arm was pulled back until the rope was taut, and a heavy stone ball was loaded in a cradle. When the rope was released, the arm quickly swung up and threw the stone toward a target. The ballista, which was similar

Archers in this stone relief stand behind a battering ram as it breaks down a castle's wall.

Giant Slingshots

The ancient Romans used a siege engine called the onager. Its name came from the kicking action of the machine, which resembled the violent kick of an onager, a type of wild donkey. Onagers worked much like giant slingshots, flinging huge stones at enemy targets. They were large wooden machines that measured more than 6 feet (1.8m) tall and weighed hundreds of pounds.

Roman soldiers prepare to fire a ballista. The ballista could throw rocks as well as spears.

to a catapult, resembled a giant crossbow and was capable of throwing spears as well as stones. In ancient Greece and Rome, the ropes of ballistae were often made from women's hair. It was considered patriotic for women to grow long hair specifically for use in an army's ballistae.

Siege engines were crude when compared with the sophisticated war machines of today. In ancient times, however, they were powerful weapons that helped armies overtake and defeat their enemies.

A modern reconstruction of a Roman ballista.

An illustration shows workers making cannons. Used with gunpowder, the cannon was the world's first explosive weapon.

The Power of Gunpowder

Machines of war became even more deadly after people learned to make gunpowder, the first known explosive. The substance is believed to have been created in China in about A.D. 500, and it was first used to make firecrackers. Chinese people set off firecrackers during religious rituals and festivals in the hope of scaring away evil spirits. Later, they used gunpowder to make simple rockets out of bamboo shoots.

News of the explosive substance spread throughout the world. By about 1200, gunpowder had made its way to Europe. Scientists created improved formulas, and by the fourteenth century, the world's first explosive weapon—the cannon—had been developed.

A cannon consisted of a bronze or iron tube that was closed at one end. It had a small hole, known as a "touchhole," drilled through the tube. An operator loaded gunpowder into the open end and then inserted a cannonball so it rested against the gunpowder. Using a red-hot iron or a burning match, the operator lit the gunpowder through the touchhole. This caused a large enough explosion to shoot the cannonball out of the open end of the cannon.

How Gunpowder Works

Gunpowder is made of potassium nitrate, charcoal, and sulfur. When a spark or flame is applied to it, gases begin to form and heat energy is produced. As the heat builds up, the gases expand rapidly, causing an explosion. If this process occurs inside an open container, such as a cannon, the explosive force can shoot an object out of the open end. Bombs work the same way, but they are created inside closed containers. As a result, extreme pressure builds up and the result is a violent explosion.

One type of cannon was a shoulder-mounted weapon called a hand gonne. These small cannons were loud, but not very powerful, and could only shoot about 30 yards (27m). Larger cannons had much more power. They were often mounted on heavy wooden frames with wheels, so they could be moved from place to place. The biggest and most powerful cannon was known as the bombard. It measured up to 15 feet (5m) long and weighed as much as 16 tons (15MT). Bombards fired stones or iron balls that weighed hundreds of pounds.

Gunpowder began as a substance that did little more than cause firecrackers to crackle and snap. When people learned how to use it in cannons, weapons became powerful explosive devices.

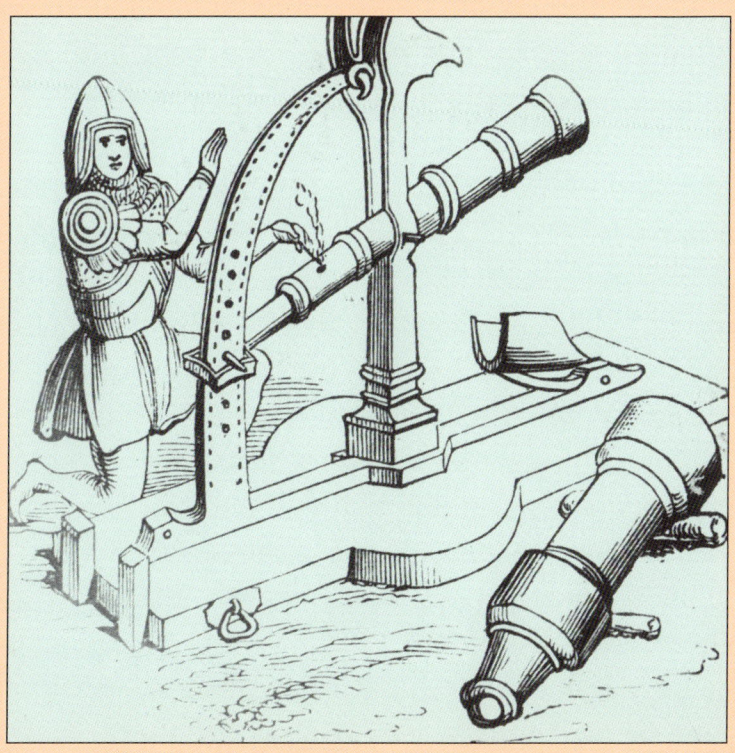

A 14th-century soldier prepares to light the touchhole of an early cannon.

The musket's matchlock firing system required the shooter to use a fuse to light the touchhole.

Handguns and Rifles

Once people were aware of gunpowder's potential, they began to make guns. A firing system known as the matchlock was developed in the 1400s, and it partially automated the ignition, or lighting, process. When an operator pulled a triggerlike device called a serpentine, a burning fuse was delivered to the touchhole. This led to the creation of the musket, a gun that became an important weapon of war. There were still problems with the matchlock system, however. A lighted match was needed to ignite gunpowder, so guns were useless in rainy weather. Also, a soldier could not sneak up on an enemy at night because the glowing flame would give him away. What was needed was an ignition system that did not require a match.

The wheel lock, which created a fire inside a gun, solved the problem. When an operator pulled the trigger, a rough-edged steel wheel rotated against a piece of hard mineral known as iron pyrite. This created a shower of sparks that ignited the gunpowder. An even better device, the flintlock, was developed in the 1620s. The flintlock was safer and less complicated to operate than the wheel lock. It led to the introduction of a variety of guns, including flintlock rifles and pistols.

In the early 1800s, a firing system known as the percussion cap was introduced. This device allowed

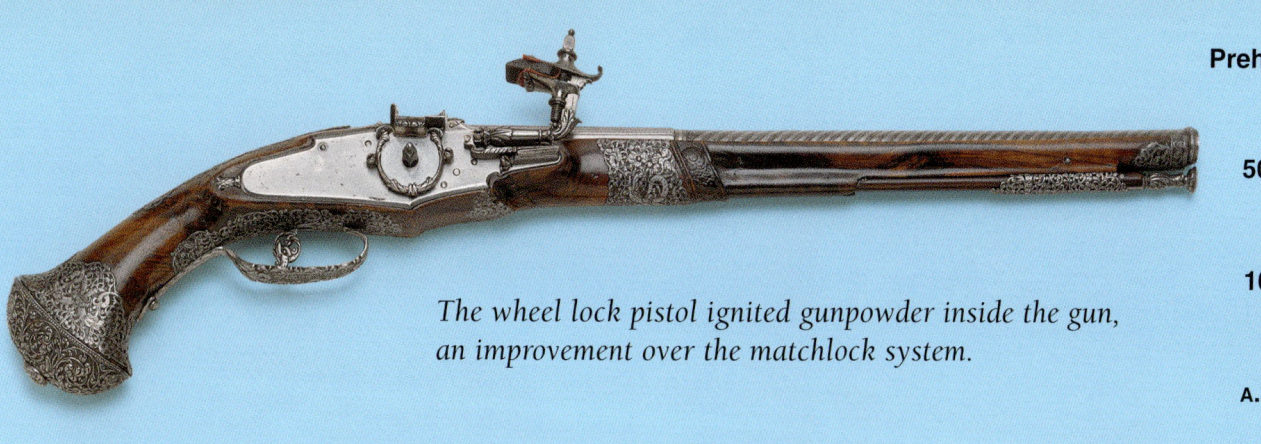

The wheel lock pistol ignited gunpowder inside the gun, an improvement over the matchlock system.

gunpowder to be ignited by the strike of a hammer rather than a spark, so it was much safer and more reliable. Later, compact cartridges were developed that held an igniter, gunpowder, and a bullet all in one piece.

From ignition systems that required an open flame to self-igniting cartridges, guns changed dramatically from the fifteenth to the nineteenth centuries. With each new feature, they became safer to use, more accurate, and more reliable.

Repeat-Firing Guns

One drawback of early guns was that they could be fired only one time. They had to be reloaded before firing again. When self-contained ammunition was introduced, guns could be shot more than once from a single loading. One of the first guns to feature this new device was the Winchester repeating rifle, which was introduced in 1849. It had a tube, called a magazine, attached to the barrel. The magazine held additional cartridges that could be loaded and fired simply by moving a lever.

These American Civil War rifles used percussion caps, a safer and more reliable firing system than earlier devices.

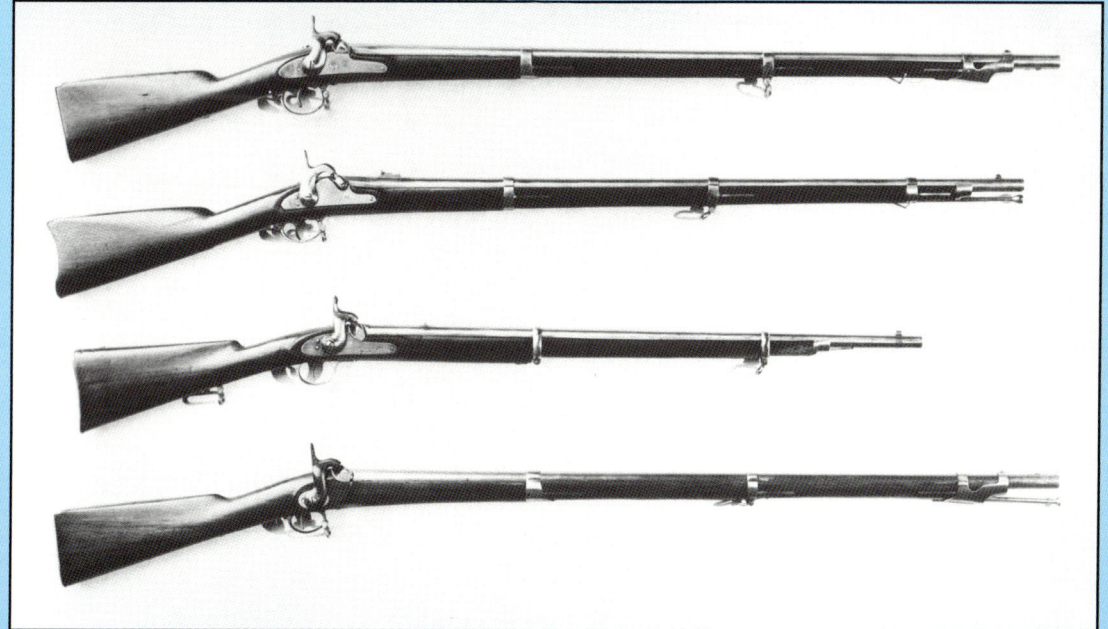

15

Alfred Nobel created dynamite by combining nitroglycerin with silica.

Powerful Explosives

In 1846, a substance was invented that was far more powerful than gunpowder. Called nitroglycerin, it was created by an Italian chemist named Ascanio Sobrero. To make nitroglycerin, Sobrero combined nitric acid, sulfuric acid, and glycerol. His early formula was extremely dangerous—even the lightest shock could cause it to explode. Sobrero found this out when an explosion badly scarred his face. After that, he considered the substance far too dangerous to be of any practical use, and he said he was ashamed to have created it.

Swedish chemist Alfred Nobel believed that nitroglycerin had great potential, and he studied it extensively. His goal was to create a safer substance that could be used in controlled situations, such as blasting rock for mining operations. After performing experiments, Nobel produced a nitroglycerin formula that was safe when handled properly.

Buried Explosives

Land mines are bombs that are buried in the ground. They are designed to explode when pressure is applied, such as when someone steps on or drives over one. Many land mines cannot be seen because the trigger mechanism is covered with dirt or grass. They pose grave danger to anyone who goes near them. There are more than 100 million land mines buried in 70 countries around the world. Since 1975, land mines have killed or maimed more than 1 million people.

Nobel later found that when he combined nitroglycerin with an element called silica, the mixture changed from a liquid into a paste. He kneaded the paste and shaped it into rods that could be inserted into drilled holes. He named his creation dynamite after the Greek word for "power."

During the same period, a German chemist named Joseph Wilbrand created trinitrotoluene, better known as TNT. It was also an explosive substance, but 30 years passed before anyone realized its power. By the early 1900s, TNT's explosive powers had been discovered, and it was found to be even safer to handle than dynamite. From that point on, TNT became the main ingredient in the deadliest weapon yet: the bomb.

Until the early nineteenth century, the only known explosive was gunpowder. With the development of nitroglycerin, dynamite, and TNT, explosive materials became safer to use, more powerful, and more deadly than any substance on Earth.

Above: Men load drill holes with dynamite in 1912.
Below: The explosive land mines in these boxes will be destroyed. Experts believe there are 100 million land mines buried in 70 countries throughout the world.

The British warship H.M.S. Dreadnought's design was so revolutionary that all later battleships were called dreadnoughts.

Weapons at Sea

People used the guns and explosives developed in the nineteenth century on the sea as well as on land. Battleships, clad with iron and heavily armed, were important weapons of war. The massive ships featured guns mounted on turrets. This allowed the guns to revolve in a circle as they were fired. They could hit targets many miles away. The most advanced battleship of its time was the British ship *Dreadnought*, which was built in 1906. This mighty vessel was plated with armor so thick that no guns could penetrate it. It had five turrets mounted with ten guns, each of which measured 1 foot (31cm) across—more firepower than any other ship on the sea. The *Dreadnought* was so revolutionary that all battleships built after it were called dreadnoughts.

Another way to launch weapons at sea was from submarines, which came into use during the early 1900s. In World War I, German submarines called U-boats were armed with torpedoes. These weapons consisted of long metal casings with an explosive charge at the tip. They

The First Combat Submarine

The first submarine used in combat was built by American inventor David Bushnell. On September 6, 1776, Bushnell's *American Turtle* carried an underwater bomb to the British ship *HMS Eagle*, which was anchored in New York harbor. Although the submarine failed to destroy its target, the British saw it as a threat and moved their fleet out of the harbor.

were powered by small electric motors. Torpedoes were fired at enemy ships, striking them below the water level, which caused them to sink. Submarines also were fitted with deck guns, which allowed them to fire at enemy ships when they surfaced.

On March 18, 1915, the *Dreadnought* rammed and sank a German U-boat in the North Sea, becoming the first battleship to sink a submarine. A few months later, the British luxury ship *Lusitania* was struck by a torpedo launched by a German U-boat. The *Lusitania* exploded and sank in just eighteen minutes. Nearly 1,200 people died, including almost 100 children.

From huge guns mounted on battleship turrets to torpedoes fired from submarines, weapons used at sea were as powerful and deadly as those used on land.

Modern Warships

Ironclad battleships remained the most important wartime sea vessels until the development of aircraft carriers in the 1920s. These ships have flat top decks that serve as a takeoff and landing site for fighter planes.

A painting shows the Lusitania *sinking after being struck by a torpedo launched from a German U-boat.*

19

Armored Vehicles

Another weapon put to use during World War I was the armored tank. The British developed this weapon because they believed armored land vehicles could be as effective as ironclad battleships. In 1915, British engineers built the first land ship, which they nicknamed *Mother*. They went on to build a fleet of more than one hundred land ships.

A British tank crosses a trench in World War I. The British developed the tank as the land equivalent of the battleship.

Each vehicle weighed 28 tons (25MT) and was covered with heavy armor. Machine guns were mounted on side platforms. The British wanted to keep the new weapons a secret from their enemies (the Germans), so they packed the vehicles in crates marked "tanks" and shipped them to battle zones in France. From that point on, all armored vehicles became known as tanks.

The British tanks were first used in the Battle of the Somme, which took place in northern France in September 1916. Forty-nine of the vehicles were used in the battle, but the results were disappointing. The

Antitank weapons

In 1942, the U.S. Army introduced a weapon called the bazooka. It was a long metal tube that fired a small rocket and was capable of destroying armored vehicles. Today, Light Antitank Weapons (LAWs) are used instead. They are similar to bazookas but use much more powerful explosives that can destroy tanks 1,000 yards (914.4m) away.

20

Ancient Armored Vehicles

In ancient times, one of the deadliest weapons used by Roman armies was the chariot. These two-wheeled, horse-drawn carriages were plated with heavy metal armor. A skilled chariot driver could crash through enemy lines while bowmen shot arrows at close range.

tanks moved less than 1 mile (1.6km) per hour, and some broke down in enemy territory. This made the vehicles an easy target for Germans with machine guns, and the bullets were able to pierce the tanks' protective armor. In spite of their failure to achieve victory in the battle, however, the British could see that tanks had tremendous potential.

As the war progressed, the British continued making improvements to tanks. Tougher armor was added, and engines were made more powerful so the vehicles could move faster. In a battle held in November 1917, the British used 400 tanks. They successfully crossed the German lines and captured thousands of enemy soldiers.

Tanks were designed to be like ironclad battleships that operated on land. With their heavy armor and powerful guns, the fighting vehicles penetrated enemy lines like no weapons ever used before.

British tanks advance with troops during World War I. Design improvements made the tanks tougher and more powerful.

War Planes

Weapons used during World War I were fired not only from land and sea, they were also fired from the air. Airplanes had been invented more than a decade before by the Wright brothers, but they had not been used as weapons. Then in 1911, Italy became the first country to drop a bomb from an aircraft. In a battle with Turkish troops over Libya, the Italians dropped 4-pound (1.492kg) explosive grenades from a single-winged aircraft known as a monoplane.

Three years after Italy's attack, World War I began. Other countries also had bomber aircraft, but very few of the planes existed. Also, the early aircraft were not very powerful. The first British bombers flew at just 72 miles (116km) per hour, and they could stay aloft for no more than three hours.

Throughout the war, airplanes were greatly improved. They reached speeds of more than 100 miles (161km) per hour and could stay in the air for up to eight hours. These aircraft were able to drop bombs, and some also dropped steel arrows called flechettes. In addition to being armed with bombs, airplanes had machine guns mounted on the wings so they could shoot at enemies in the air and on the ground.

Although difficult to fly, the British Sopwith Camel was an effective fighter aircraft in World War I.

A British Sopwith Camel fires at German biplanes. The Sopwith Camel's two machine guns were deadly to enemy aircraft.

The most successful fighter aircraft of World War I was a British plane known as the Sopwith Camel. This small, lightweight biplane was considered extremely deadly. It had two machine guns enclosed in a hump, which gave the Camel its name. Over the course of the war, Sopwith Camels shot down more enemy aircraft than any other fighter plane. As effective as the planes were, however, they were extremely tricky to fly. More pilots died while learning to fly them than using them in combat.

From the first airplanes to the famous Sopwith Camels, much of World War I's firepower came from the skies above.

Attacked by Balloons!

On August 22, 1849, the Austrian army launched 200 unmanned balloons over Venice, Italy. The balloons were made of paper and had copper wires attached that triggered fuses for bombs. The balloons dropped 33-pound (15kg) bombs designed to explode and catch fire to the surrounding area. No great damage was done by the Austrian bombs, although one of the charges exploded in an area known as St. Mark's Square. When an unexpected shift of the wind drove some of them back toward the soldiers who launched them, the bomber balloons were abandoned.

Chemical Weapons

Soldiers cling to each other after being blinded by mustard gas, a chemical weapon used in World War I.

By the end of World War I, a frightening and deadly weapon had been unleashed—one made from poisonous chemicals. Several countries used chemical weapons during the war. France released grenades containing a lethal chemical called turpinite against German soldiers. Germany retaliated by launching the world's first large-scale chemical attack in Ypres, Belgium, against French and British troops. Using large gas cylinders, the Germans released 160 tons (145MT) of deadly chlorine gas into the air. By the time the soldiers noticed the greenish yellow cloud wafting toward them in the breeze, it was too late. The poisonous gas burned the soldiers' lungs, killing thousands and severely injuring thousands more.

Another type of chemical weapon used during the war was mustard gas. It burned and destroyed the skin, eyes, and lungs of anyone who came into contact with it. A third type, called phosgene, was also used and was ten times more deadly than chlorine gas.

A Deadly Terrorist Attack

One of the deadliest chemical weapons is called sarin, a colorless and odorless poisonous gas. Sarin was originally developed for use as a pesticide. In 1995, a group of terrorists released containers of the deadly gas on a subway in Tokyo, Japan. The attack killed 12 people and injured nearly 6,000.

The horrific effects of chemical weapons brought many countries together after the war in an effort to ban them. In 1925, 38 nations signed a treaty known as the Geneva Protocol. It did not end production of chemical weapons, but it helped discourage countries from developing them.

During World War II, Germany developed chemical weapons but never used them. In 1945, about 300,000 tons (272,155MT) of weapons filled with mustard gas and other poisonous chemicals were discovered in German arsenals. To dispose of them, Allied forces (Great Britain and the former Soviet Union) dumped thousands of tons of the chemicals into the Baltic Sea.

Chemical weapons have not been used in widespread attacks since World War I. Due to their capacity to cause massive injury and death, they are still considered some of the world's most dangerous weapons.

Iraqi Massacre

In 1988, Iraq's former president, Saddam Hussein, launched a poisonous gas attack against a group of Iraqis known as Kurds. It was one of the largest and most deadly civilian attacks in history. Nearly 1,300 Kurdish villages were destroyed, and thousands of people were killed.

A soldier in 1918 experiences the toxic effects of phosgene gas, ten times deadlier than chlorine gas.

25

Nuclear Weapons

As deadly as chemical weapons were, there was something that proved to be far more dangerous and destructive: the nuclear bomb. During World War II, scientists studied radioactive atoms. They learned how to split atoms by bombarding them with neutrons in a process known as fission. Scientists called the development "harnessing the basic power of the universe." The United States used fission to develop the world's first atomic, or nuclear, bomb.

By 1945, the war between the United States and Japan had gone on for nearly four years. The United States called for Japan's surrender, but the Japanese would not agree unless their emperor would be allowed to remain in power. In an effort to force their surrender, the president of the United States decided to use nuclear bombs against them. On August 6, 1945, an aircraft called *Enola Gay* dropped the world's first nuclear bomb on Hiroshima, Japan. The bomb known as *Little Boy* was extremely powerful—equal to more than 20,000 tons (18,142MT) of TNT. Three days later, another U.S. airplane dropped a second

A mushroom cloud rises on August 6, 1945, from the explosion of the nuclear bomb Little Boy *over Hiroshima, Japan.*

nuclear bomb, called *Fat Man*, on Nagasaki, Japan. The two bombings virtually destroyed the two Japanese cities and killed more than 200,000 people.

In 1949, the Soviet Union developed and tested its own atomic bomb. The United States followed by creating a hydrogen bomb, which used a type of nuclear reaction called fusion. Hydrogen bombs were about 700 times more powerful than atomic bombs. Since that time, several other countries have also developed nuclear technology. Not since the bombings of Hiroshima and Nagasaki, however, has the nuclear bomb been used against another country.

Of all the weapons ever made, only nuclear weapons can harness the power of the universe. They have the capability to destroy the world.

Lost Nuclear Weapons

In 1966, an American bomber carrying four hydrogen bombs collided with another aircraft over Spain. Three of the bombs fell to the ground without exploding, and one fell into the Mediterranean Sea. More than 3,000 people worked on the search and clean-up operation, which took 80 days. Wreckage from the disaster covered almost 100 square miles (259sq km) of land and water.

The United States dropped a second nuclear bomb on Nagasaki, virtually destroying the city.

27

Biological Weapons

Unlike chemical or nuclear weapons, biological weapons have never been used in warfare. Their potential, however, is of great concern. They are designed to spread deadly bacteria or viruses. That means they have the capacity to kill millions of people.

Biological weapons are created with living microorganisms known as biological agents. There are three groups of these agents: bacteria, viruses, and toxins. They can be released into the air or added to drinking water or food. They can also be added to soil, which in turn can contaminate crops.

One of the most dangerous biological agents is anthrax, which is a serious bacterial disease. Another is smallpox, a disease that is highly contagious. A particularly deadly biological toxin is botulinum, a type of bacteria that is so deadly a mere one-billionth of an ounce can be lethal. Another potential agent is the plague, a deadly disease that can lead to shock and death in a matter of days.

There have been only two known attacks with biological weapons, and both occurred in the United States. The first happened in 1984, when members of a religious cult

The dangerous biological agent anthrax is shown under a microscope. Biological weapons can kill millions of people.

Early Biological Weapons

The first known use of biological weapons was in China in the mid–fourteenth century. Mongols threw diseased bodies over their enemies' walls in an effort to frighten them away. In colonial America, British troops delivered blankets from a hospital to American Indians. The blankets had been used on patients with smallpox, and the British were hoping to infect the Indians with the deadly disease.

At least five deaths occurred in 2001 when letters containing anthrax spores were sent through this U.S. Postal office.

placed salmonella bacteria in restaurant salad bars. Seven hundred people became ill, but there were no deaths. The second attack, which occurred in 2001, was launched through the U.S. Postal Service. Letters containing a white powder were mailed to targeted people. The powder contained anthrax spores, and the attack caused 22 cases of infection. At least 5 of the people who had inhaled the anthrax died.

Unlike other weapons, biological weapons do not make loud noises or create fiery explosions. In fact, they cannot be seen or heard at all. They are silent weapons that spread bacteria, toxins, and disease—their capacity to kill is devastating.

Weapons of Mass Destruction

Chemical, biological, and nuclear weapons have the capacity to kill thousands or even millions of people. That is why they are called weapons of mass destruction.

Workers in protective clothing check an area for evidence of anthrax contamination.

29

Weapons in a High-Tech World

A U.S. F-18 fighter jet carries a full load of missiles and bombs. Many such weapons today are guided by computers.

Today there are many highly sophisticated weapons that are no longer dependent on the human eye. Many weapons are now controlled by computers. One example is the cruise missile, which can be launched from aircraft, ships, or submarines hundreds of miles from a target. By following an electronic map in an internal computer, a cruise missile can strike within a few feet of the target it was programmed to destroy.

Another example of high-tech weaponry is the smart bomb. Unlike bombs of the past, known as dumb bombs, that fell wherever gravity took them, smart bombs can be directed toward a precise target. These bombs read laser signals sent to them from the ground, which guide them to fly into and destroy targets.

America's stealth bomber is a unique aircraft that resembles a giant boomerang. Unlike other planes that have two wings, the stealth has one wing that is 172 feet (52.43m) wide. It can travel for nearly 7,000 miles (11,265km) before refueling, and its design makes it invisible to radar. The stealth bomber is capable of traveling anywhere in the world on very short notice.

One high-tech weapon is the Predator, which is a type of Unmanned Aerial Vehicle (UAV). The Predator is an unusual aircraft because there is no pilot inside it. Instead, crews on the ground fly it by

High-Tech Missile

The Patriot missile is an amazing defensive weapon that is designed to detect, target, and destroy incoming missiles. Using built-in radar and a guidance computer, it can track missiles 50 miles (80km) away, determine their speed, and then destroy them before they reach the intended target.

Electrical Weapons

Instead of bullets, stun guns shoot metal prongs with attached wires. The prongs stick to a person's skin or clothing. An electrical charge is then sent through the wires, causing an electrical shock. The weapon's effects are temporary, and the person suffers no permanent injuries.

The stealth bomber is invisible to radar and can fly great distances before refueling.

remote control. This is valuable during wartime because the aircraft can fly into combat zones without risking the lives of its crew, who are often located many miles away. The Predator is capable of flying long distances and can stay in the air for as long as twenty-four hours. Its onboard cameras take pictures and send them back to the base, where they are used to monitor enemy locations and activity. Sophisticated radar equipment allows the plane to "see" through thick haze, clouds, or smoke. Also, the Predator has the ability to fire laser-guided missiles at targets on the ground.

From the stones and spears of ancient times to today's computer-controlled aircraft, people have always considered weapons essential for survival. In the future, even more powerful weapons will likely be created. The challenge for humans will be to figure out how to use these weapons without destroying themselves in the process.

The unmanned Predator can gather information with its onboard cameras and fire missiles at enemy targets.

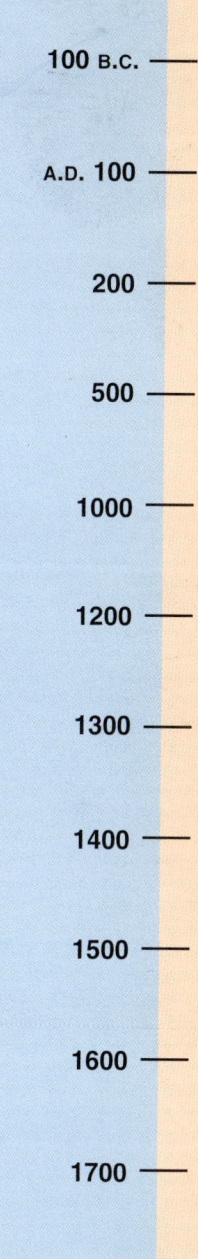

31

Glossary

agent: A substance that causes some sort of change.
bacteria: Microscopic organisms that carry and spread disease.
casting: The process of pouring melted metal into molds to make weapons or other objects.
chain mail: Armor made with thousands of iron rings tightly linked to form metal fabric.
fission: A nuclear reaction caused by splitting atoms into pieces.
flint: A hard, brittle stone.
fusion: The process of combining the nuclei of atoms to form much larger nuclei.
magazine: A tube or box attached to a firearm that holds ammunition.
projectile: Any object that is shot or hurled through the air.
serpentine: A triggerlike device used to fire a gun.
siege engines: Ancient weapons such as battering rams and catapults that were used to destroy and capture fortified cities.
sinew: Animal tendon.
touchhole: A small hole drilled into the end of a cannon used to ignite gunpowder.
trinitrotoluene: An explosive substance used to make bombs, commonly called TNT.

About the Authors

Peggy J. Parks holds a bachelor of science degree from Aquinas College in Grand Rapids, Michigan, where she graduated magna cum laude. She is a freelance writer who has written more than 30 titles for Thomson Gale's KidHaven Press, Blackbirch Press, and Lucent Books.

For More Information

Books

Michèle Byam, *Arms & Armor*. New York: DK, 2004.
Michael and Gladys Green, *Assault Amphibian Vehicles: The AAVs*. Mankato, MN: Capstone, 2004.
Milton Meltzer, *Weapons & Warfare: From the Stone Age to the Space Age*. New York: HarperCollins, 1996.
Rob Staeger, *Native American Tools and Weapons*. Philadelphia, PA: Mason Crest, 2003.

Periodicals

Jonathan Blum, "The Power of Flight," *Scholastic Action*, December 8, 2003.
Jamie Kiffel, "The Legend of the Samurai," *National Geographic Kids*, January/February 2004.
Richard W. Slatta, "South America's Cowboys," *Cowboys & Indians*, May 1997.

Web Sites

History for Kids—Ancient War (www.historyforkids.org/learn/war). A site designed for young people that explains why there have been wars throughout history and how those wars were fought.
How Stuff Works (www.howstuffworks.com). Excellent information about weapons can be found in this site, including different types of guns, explosives, missiles, and biological and chemical weapons.
ThinkQuest—Weapons (www.thinkquest.org/library/cat_show.html?cat_id=171). An excellent portal site with links to a number of weapons-related Web sites. All of the Web sties were created by students and have won ThinkQuest awards for excellence.

Index

anthrax, 28, 29
armor, 6, 8, 18, 20, 21
armored vehicles, 20–21
arrows, 6–7
atomic bomb, 26

battering ram, 10
battleships, 18–19
bazooka, 20
biological weapons, 28–29
bombs, 13, 17, 18, 22, 23, 26, 27, 30
boomerangs, 5
bow and arrow, 6–7
bronze, 8–9

cannon, 12–13
catapult, 10
chemical weapons, 24–25
clubs, 4
copper, 8
cruise missile, 30

daggers, 8
dynamite, 17

explosives, 12, 16–17

fighter planes, 22–23
flint, 7, 8

grenades, 22, 24
gunpowder, 12–13, 15

handguns, 14–15
high-tech weapons, 30–31

landmines, 16

machine guns, 20, 21, 22, 23
metal, 8–9

nuclear weapons, 26–27

planes, 22–23, 30–31

rifles, 14–15
rocks, 4

ships, 18–19
siege engines, 10–11
slingshot, 5, 10
smart bomb, 30
Sopwith Camel, 23
spears, 4–5
stealth bomber, 30–31
submarines, 18–19
swords, 9

tanks, 20–21
torpedoes, 18–19

U-boats, 18–19

warships, 18–19
weapons of mass destruction, 29